MINERALS

Rebecca Woodbury, Ph.D., M.Ed.

Gravitas Publications Inc.

MINERALS

Illustrations: Janet Moneymaker

Copyright © 2025 by Rebecca Woodbury, Ph.D., M.Ed.

Minerals
ISBN 978-1-950415-31-1

Published by Gravitas Publications Inc.
Imprint: Real Science-4-Kids
www.gravitaspublications.com
www.realscience4kids.com

RS4K
Photo credits: Cover & Title Pg: Demetrio, AdobeStock; Above, Didier Descouens, CC BY SA 4.0; P.11. 1) Parent Géry, CC BY SA 3.0; 2-4) Rob Lavinsky, iRocks.com; CC BY SA 3.0; 5) Public Domain; P.15. 1) JJ Harrison (www.jjharrison.com.au) CC BY SA 2.5; 2) Alexander Potapov, AdobeStock; 3) Sebastian, AdobeStock; 4) Didier Descouens, CC BY SA 4.0; P.17. 1) Moha112100, CC BY SA 3.0; 2&3) Public Domain; 4) Tim Evanson from Washington, D.C., USA, CC BY SA 2.0; P.19. 1) Sabrinna Ringquist on Unsplash; 2) byjeng, AdobeStock; 3) Sapphiredge, CC BY SA 3.0; 4) Philippe Van-doninck on Unsplash; 5) Silverborders, CC BY SA 3.0; 6) Anastasia Tsarskaya, AdobeStock; 7) vitaly tiagunov, AdobeStock; 8) Elena, AdobeStock; P.21. grach_a, AdobeStock

What are rocks and dirt made of?

I like dirt!

Rocks and dirt are made of **minerals.**

Minerals form when **magma** deep inside the Earth cools very slowly.

Magma is very hot, **molten** (melted) material.

Did you know that Earth is hot in the middle?

No. I did not.

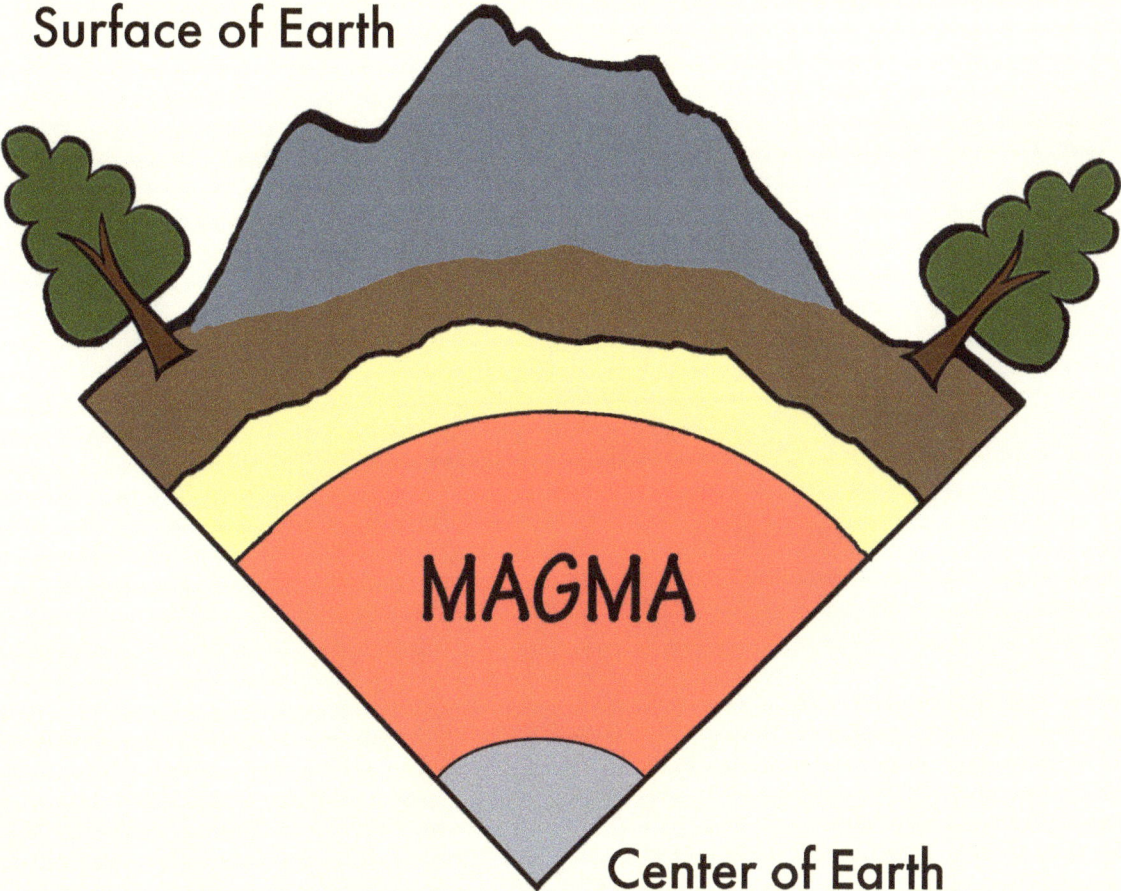

Surface of Earth

MAGMA

Center of Earth

When magma cools slowly, the **molecules** in it have a chance to line up in an orderly way.

When this happens, a **mineral** is formed.

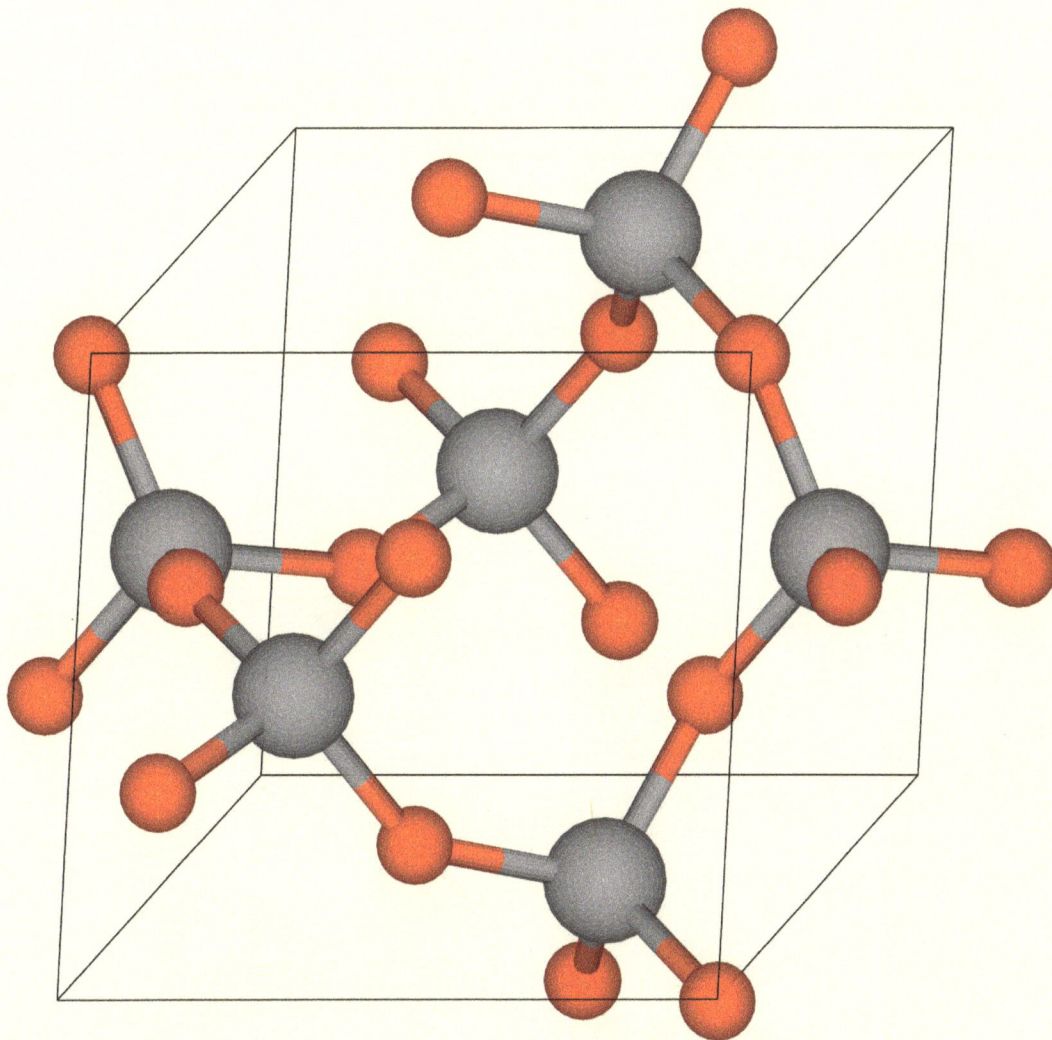

Molecules are lined up in an orderly way.

Review: MOLECULES

Molecules are made when **atoms link** together.

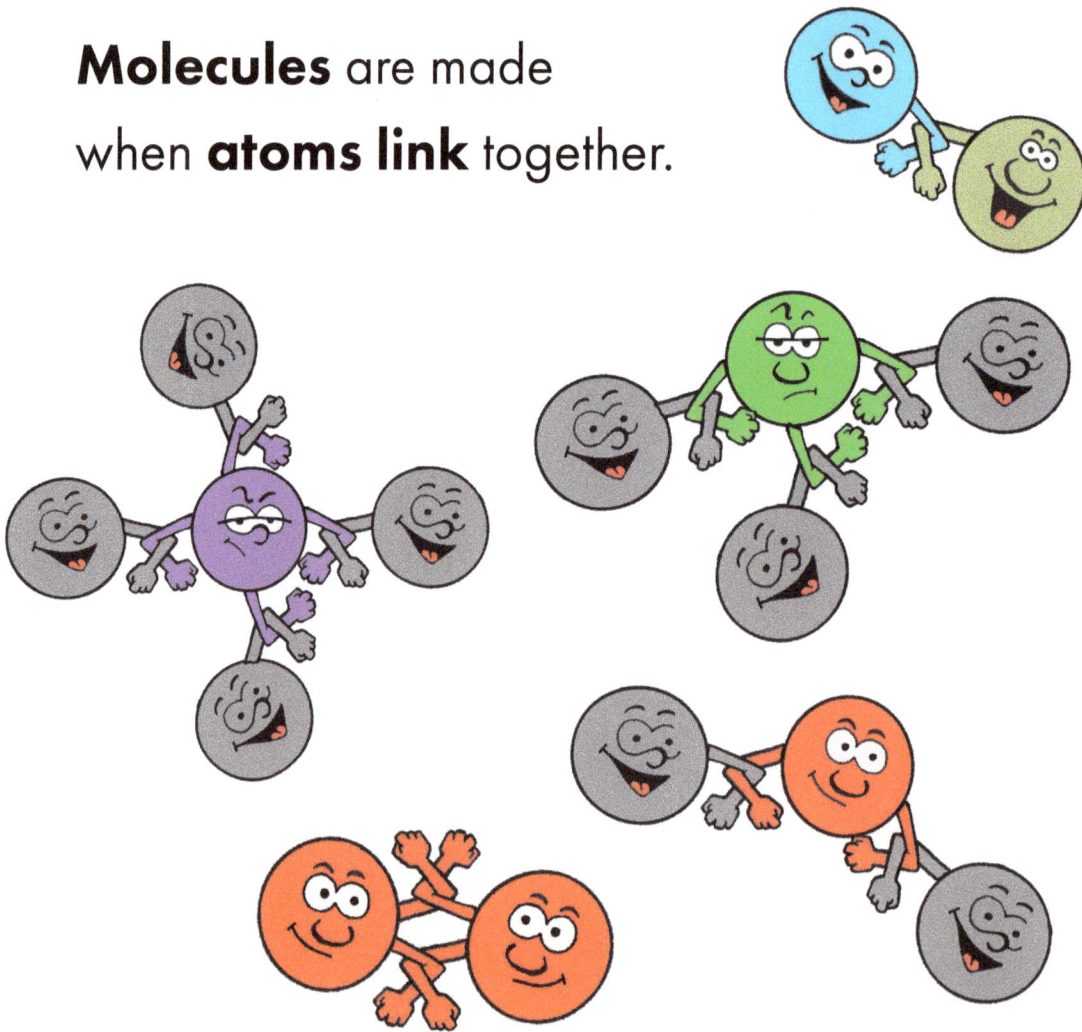

Review: ATOMS

- **Atoms** are tiny building blocks that can link together.

- **Atoms** make up everything we see, touch, taste, and smell.

Because minerals are made of molecules that are lined up in an orderly way, they can form **crystals**. Each type of mineral has crystals that are a specific shape.

Page 11 mineral photos:
1. Emerald and quartz; 2. Beryl; 3. Calcite;
4. Beryl, quartz, morganite; 5. Pyrite

1

2

3

4

5

Minerals made of **silicon** atoms and **oxygen** atoms are called **silicates**.

Do you think these atoms are in other things too?

Must be.

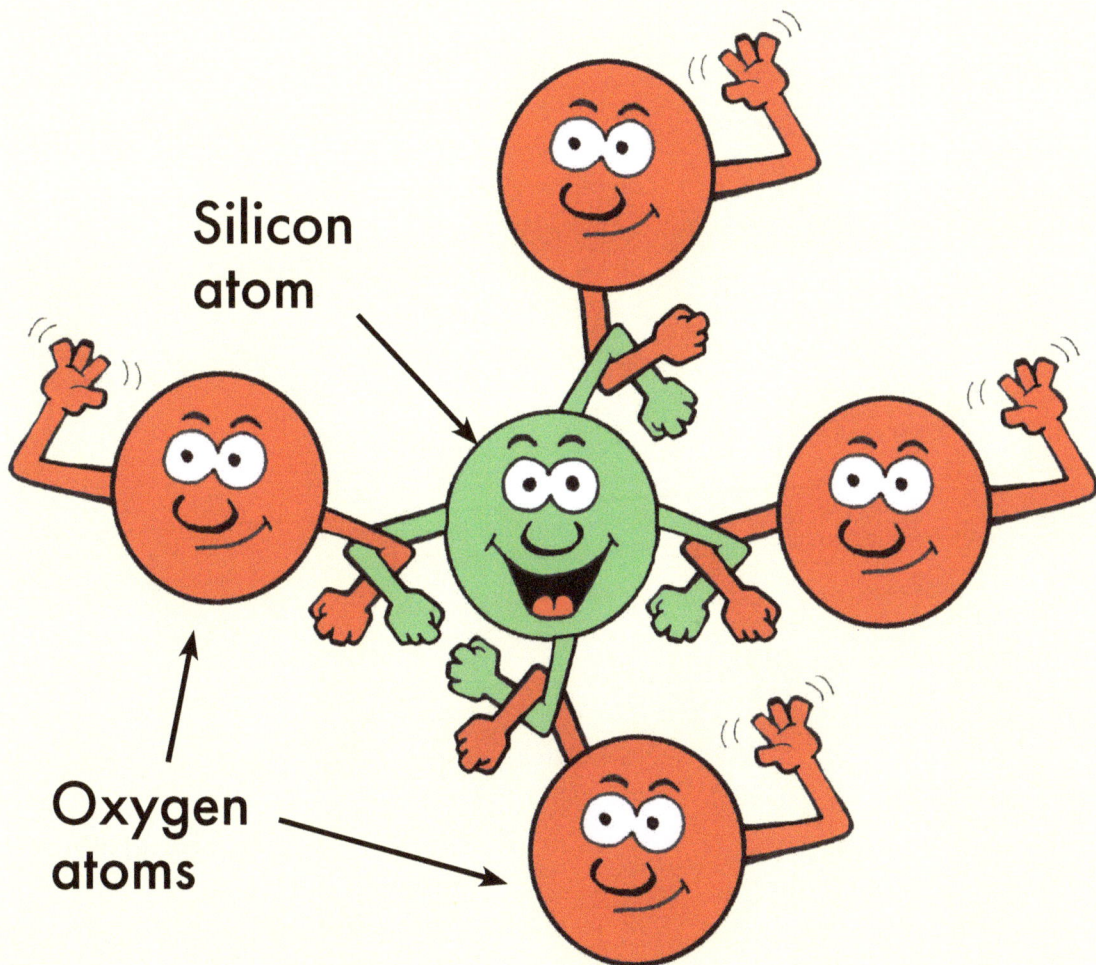

Silicon atom

Oxygen atoms

Quartz is a silicate mineral made of silicon atoms and oxygen atoms.

Wow! Quartz has cool shapes and colors!

1

2

4

3

Minerals can come in different colors. Small amounts of different atoms in the crystals create the different colors.

Garnet Contains iron

Tourmaline Contains manganese

Emerald Contains chromium

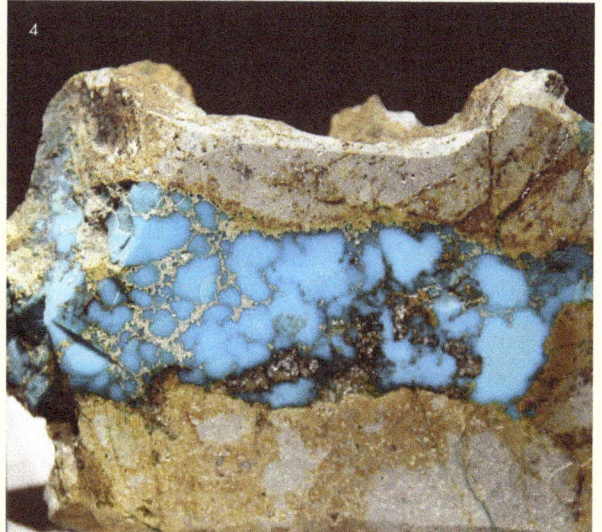

Turquoise Contains copper

Sometimes these colorful minerals are cut and polished to make jewelry.

We are so lucky to have minerals!

Yes!

Minerals are among Earth's most wonderful treasures.

Treasure!

How to say science words

atom (AA-tuhm)

chromium (KROH-mee-uhm)

copper (KAH-puhr)

crystal (KRIS-tuhl)

emerald (EM-ruhld)

garnet (GAHR-nuht)

geologist (jee-AH-luh-jist)

iron (IY-uhrn)

magma (MAAG-muh)

manganese (MAAN-guh-neez)

mineral (MIN-ruhl)

molecule (MAH-lih-kyool)

molten (MOHL-tuhn)

oxygen (AHK-sih-juhn)

quartz (KWAWRTS)

silicate (SIH-luh-kayt)

silicon (SIH-luh-kahn)

tourmaline (TUHR-muh-leen)

turquoise (TUHR-kwoyz)

www.ingramcontent.com/pod-product-compliance
Lightning Source LLC
Chambersburg PA
CBHW040150200326
41520CB00028B/7559